TELEPHONE BOXES

JU

Gavin Stamp

TELEPHONE BOXES

Chatto & Windus
LONDON

Published in 1989 by
Chatto & Windus Ltd
30 Bedford Square
London WC1B 3SG

A CIP catalogue record for this book
is available from the British Library

ISBN 0 7011 3366 X

Photoset and printed in Great Britain by
Redwood Burn Limited, Trowbridge, Wiltshire

CONTENTS

ACKNOWLEDGEMENTS

I have learned much about the history of the telephone kiosk from my colleagues in the Thirties Society and, in particular, from those two stalwart heroes in the struggle against British Telecom's campaign of destruction: Philip Davies and Dominic West, conservation officers respectively for the City of Westminster and the London Borough of Camden. I must also record my great debt to the librarians at the Resource Centre of the Telecom Technology Showcase, who most generously made archival material and historical illustrations available to me. I have also enjoyed assistance from Neil Burton, Kate Emms of English Heritage, Jonathan Glancey, Roger Last of the BBC, Nicholas Long, Bruce Martin, Alan Powers, John Martin Robinson, Douglas Scott, David Walker, Lesslie Watson and the staff of the RIBA Drawings Collection. Unless otherwise attributed, all photographs were taken by the author.

G.M.S.
December 1988

TELEPHONE BOXES

Kiosks are architecture; boxes and booths are not. 'Kiosk ... 1625 [... Turk. *kiushk* pavilion – Pers. *guš[a* pavilion]. 1. A light open pavilion or summer house, often supported by pillars; common in Turkey and Persia. 2. A light structure, resembling this, for the sale of newspapers, a bandstand, etc. 1865 ... ' So the *Shorter Oxford English Dictionary*. Kiosks are to be found all over Istanbul, for instance; they are open ornamental pavilions often intended for feasting and 'generally not intended for permanent occupancy'. Nor, of course, is a telephone kiosk, but there the resemblance ends. It was a strange process of transmogrification which made this Turkish word also serve to describe the elegant Classical cast-iron, teak and glass boxes for public telephones designed by Sir Giles Gilbert Scott which were once the ubiquitous ornament of Britain, both urban and rural, as well as other small shelters for this instrument for communication.

In Europe, kiosks first appeared in the country, as ornamental pavilions in landscaped gardens serving similar purposes to their Turkish prototypes. Interest in the Orient encouraged such conceits, and the evolution into such things as band-stands in parks was natural. With the nineteenth-century love of the exotic and the ornamental, it was not surprising that functional booths or stalls in towns should be dressed up elaborately to resemble Turkish kiosks. Such things were commonplace in streets and railway stations by the middle of Victoria's reign. The full *Oxford English Dictionary* tells us that 'kiosk' meant, secondly, 'A light ornamental structure ... used for the sale of newspapers (in France and Belgium), for a band-

stand, or for other purposes. 1865 *Daily Tel.* ... a 'kiosk' – i.e. a
place for the sale of newspapers...' That was in 1933. The 1976
Supplement reveals that this definition is 'No longer restricted to
France and Belgium' and now encompasses the 'telephone kiosk'.
This was almost a century after the things began to appear in the
streets of Britain's cities.

It is still a puzzling stylistic journey from the ornamental
columns, pointed arches and domed projecting roofs of a Turkish
kiosk to the neo-Grecian refinement of Scott's box. A clue is
suggested by one of the early public telephone boxes which stood in
High Holborn (p.29). This structure of cast iron, erected in about
1903, was octagonal in plan and was surmounted by cresting and
an ornamental domed roof. It resembled nothing so much as a
Parisian advertising drum-cum-newspaper stall. In other words, it
really was a *kiosk*. But it was exceptional. Most early public
telephones were placed in simple glazed wooden cabinets which
most resembled sentry boxes. If placed indoors, as 'silence
cabinets', they would have a flat top; if outdoors, a double-pitched
shed roof.

'Public call offices' were first authorised by the Postmaster
General in 1884. At this date, the first telephone exchanges in the
country were only five years old and there were some 13,000
telephones throughout the British Isles. The various private
telephone companies were naturally anxious to expand business and
competed to install public call offices in busy places like railway
stations, hotels and shops. However, access to 'silence cabinets' in
shops was restricted by opening hours and by the attitude of the
shopkeepers so that, after the turn of the century, the companies
began to erect free-standing external kiosks for easier use. These
were either entered by inserting coins into the lock (as on public
lavatories) and operated automatically via the operator (payment
inserted in advance) or opened and operated by an attendant. Many
of these kiosks bore the sign of a blue bell in addition to the notice
stating 'Public Telephone'. This was the symbol of the National

Telephone Company which, by 1907, had absorbed most of its rivals and operated 7,800 call offices throughout Britain.

Most of these early telephone kiosks were of the ordinary wooden sentry-box or road-mender's hut type, but there was little uniformity (p. 30). Different companies erected different types of box, while certain local authorities insisted on special designs intended to harmonise with their surroundings. Rustic huts, like garden pavilions, were particularly favoured, like those erected on the sea front at Folkestone. A most extreme case was the polygonal kiosk placed outside Blackburn in 1907. This had rustic walls of sliced logs, a pitched wooden roof and leaded-light windows of a suitably artistic pattern (p. 30).

In 1912 the General Post Office took over the control of almost all of the national telephone network (Kingston-upon-Hull alone remains independent today). In 1913 the Post Office considered producing a standardised ornamental kiosk and using the colour red, but nothing was done to achieve standardisation until after the Great War. In 1921 the Post Office began to develop the Kiosk No. 1 to meet this need. This was essentially an improved version of the ordinary wooden sentry-box type except that it was usually of reinforced concrete construction and had metal glazing bars (p. 32). It also had a pyramidal roof, ornamented with an iron finial, rising above a projecting cornice rather than a shed roof. The K 1, in several variants (some in wood), went into production in 1923. Five hundred had been ordered by March 1923, and in 1925 they were being manufactured at the rate of fifty-two a month (at a unit cost of £13). 6,300 of these were in use in 1934, but very few survive.

Aesthetically, the K 1 was not a success. It was essentially Edwardian in conception and seemed a crude and old-fashioned design for the 1920s, when there was a growing consciousness of the importance of higher standards in public design. Many local authorities disliked the kiosk and placed restrictions on its employment by the Post Office. In Eastbourne, the council insisted that K 1s placed along the sea front should be given a thatched roof.

The effect, on so narrow a structure, was top-heavy and quite grotesque.[1] (p.34). The Post Office's kiosk was particularly disliked in the London area. Some borough councils refused permission for any to be erected and this opposition, combined with objections from the Ministry of Transport, resulted in there being only 180 public telephone kiosks placed on the public highways of the capital by 1926. By 1929 there were 1,581 – but this dramatic increase was the result of an acceptable new design being produced by the GPO.

Dissatisfaction with the Kiosk No. 1 was such that in 1923 the Metropolitan Boroughs Joint Standing Committee decided to organise a competition to secure a superior design which might be more acceptable in Britain's streets. Unfortunately, none of the designs submitted was very inspired. Most were essentially more ornamental versions of the K1 and did not suggest a radical conception. A curious entry came from the elderly Arts and Crafts architect, C. F. A. Voysey. This, designed for Messrs Vickers, was Gothic in style with heraldic panels and ornamental cresting, made of cast aluminium. The body was to be made of plywood, 'enamelled bright vermilion ... the whole would be easily cleaned by turning on a hose.'[2] This, and other Gothic designs, were not highly regarded. Nor does there seem to have been an outright winner, for in December 1923 the Office of the Engineer-in-Chief of the GPO produced a design for a kiosk in cast iron which, presumably, was intended to satisfy the Metropolitan Boroughs. This was essentially an improved K1 with simplified glazing, a flattened roof of ogee profile and an ornamental panel with the monogram of the GPO below the glass on the door. The legend 'Telephone' which on the K1 was borne by a notice attached to the roof, was transferred to the top of the door(p.40).

Meanwhile, a much more radical and intelligent idea for a public call box had been produced by the Birmingham Civic Society. This had large panels of glass, flat angled piers at the four corners, and a subtly ornamented barrel vault fitting over the depressed arch of the door. The whole was to be constructed in reinforced concrete. This

unprecedented concept was approved by the Birmingham Public
Works Department and the Birmingham Advisory Art Committee
and in July 1923, a model of the design was submitted to the
Director of Telephones in London. He was evidently unimpressed,
for the Civic Society was eventually informed that the design
produced by the Office of the Engineer-in-Chief was preferred.
This, although apparently approved by the Metropolitan Boroughs
Joint Standing Committee, was not only inferior to the Birmingham
design, but was also more expensive at £35 rather than £23 per
kiosk. As the *Architects' Journal* commented in February 1924, 'no
one with any knowledge of design could feel anything but
indignation with the pattern that seems to satisfy the official mind.'[3]

However, the Council of the Birmingham Civic Society did not
let this decision go unchallenged, and asked similar bodies to use
their influence in favour of a more worthy design. In this
endeavour, the Society seems to have been successful and it was
instrumental in achieving the final happy outcome to the problem of
the telephone kiosk. The Postmaster-General was soon faced with
pressure from such bodies as the Royal Institute of British
Architects, the Town Planning Institute and the Royal Academy to
think again – as well as with the obstructiveness of many London
boroughs. Eventually, as the GPO's favoured kiosk design 'met with
general disfavour',[4] the Postmaster General decided to hold a new
competition to be organised by the Royal Fine Art Commission.
This was a wise decision, as the Commission had been very recently
established to look into aesthetic matters of public interest of
precisely this sort. Its Royal Warrant, dated 30 May 1924, stated
that 'a Commission should forthwith issue to enquire into such
questions of public amenity or of artistic importance as may be
referred to them ... and furthermore, to give advice on similar
questions when so requested by public or quasi-public bodies,
where it appears to the said Commission that their assistance would
be advantageous ... '

The new Commission recommended a limited competition. As

half of the Commissioners were architects – including Sir Edwin
Lutyens, Sir Aston Webb and Reginald Blomfield – as was the
Secretary, H. C. Bradshaw, it was not surprising that the
Commission considered that an architect was the best person to
design what was, in effect, a miniature building rather than an
ornamented sentry box. Three highly regarded practitioners were
approached: Sir Robert Lorimer, the Scottish country house
architect and the designer of the Scottish National War Memorial in
Edinburgh Castle; Sir John Burnet, another Scot, who had been
responsible for the much admired King Edward VII wing of the
British Museum; and Giles Gilbert Scott, who was English. The
general conditions given to competitors (along with the blueprint of
the K1 for reference) recommended cast-iron construction, and a
unit cost not to exceed £40.

The competitors reported back in May 1924. Lorimer offered a
fussy and ill-considered design for cast iron with a teak door. Walter
Macfarlane & Co., the Glasgow ironfounders, estimated that this
design would cost £34.10.0d each for a run of 500. Lorimer
considered that, 'as the kiosks are not to be erected in groups, the
question of their being sound-proof does not appear to be of great
importance.'[5] Burnet disagreed. His kiosk was to be double-glazed
and have double teak panelling. He also recommended that the iron
be painted the colour of oiled teak. As his comparatively simple box
was to be surmounted by a glazed semi-circular dome resembling a
lamp-shade, rising from an octagonal base, the whole was very
expensive at £50.10s.0d each for a run of 1,000. Scott's kiosk was to
be made of mild steel with, again, a door of teak. His design – much
more Classical in conception than that by the great Burnet, who had
been trained at the Ecole des Beaux-Arts in Paris – was estimated at
not over £40 each. Interestingly, Scott recommended that it be
painted silver externally and a 'greenish blue'[6] inside. The legend
'Telephone' was to be of pierced letters filled with blue glass.

Later in the year, full-sized wooden models of these three designs
were erected behind the National Gallery along with the

Birmingham Civic Society's design, now adapted for cast-iron construction and for double-glazing, and the GPO's favoured design approved by the Metropolitan Boroughs Joint Standing Committee. There can surely have been no doubt in the minds of the interested public as to which conception was by far the best and, in February 1925, it was announced that the Royal Fine Art Commission had recommended to the Postmaster General that the kiosk designed by Sir Giles Gilbert Scott, RA, was the most 'suitable for erection in busy thoroughfares of large towns'.[7]

Scott had produced a classical design of refined sophistication and timeless elegance which, for over half a century, was permitted both to enhance public spaces in towns and to become a representative symbol of British life. The Kiosk No. 2, as it was to be called by the GPO, was solid, beautiful and practical (p.42–43). It was arguably one of the very best examples of British industrial design. Yet Scott was, perhaps, an unlikely creator of such an object, for he had no training in the sphere of industrial design, nor was he by training a Classicist. The articled pupil of the church architect, Temple Moore, and the grandson of Sir Gilbert Scott, architect of the Albert Memorial and St Pancras Station, Scott was principally known as a designer of Gothic churches. His great work was the Anglican Cathedral in Liverpool, which he had won in competition in 1902 at the age of twenty-one. In July 1924, between submitting his design for a kiosk and learning that he had won the competition. Scott had been knighted at the consecration of the choir of his great work in Liverpool. But Scott was not exclusively a Gothicist. He had been influenced before the Great War by the contemporary taste for a revived Neo-Classicism and this manifested itself in several of his secular designs of the 1920s: Clare Memorial Court at Cambridge and his own house in Clarendon Place, Bayswater, as well as the telephone kiosk.

It was not at all surprising that the winning design for the telephone kiosk was Classical in inspiration. Not only was the Classical manner favoured by the leading members of the

architectural profession in Britain, but the Classical revival of the
first two decades of this century had itself encouraged an interest in
public design and civic responsibility. Both the first well-designed
standard telephone kiosk and the Royal Fine Art Commission
which chose it were products of this informed and enlightened
climate of opinion in which architects played a leading role. The
revival of interest in the architecture of the period of Wren and
Gibbs in the 1890s was a reaction to the stylistic eclecticism of mid-
Victorian architecture. Soon after the turn of the century,
'Edwardian Baroque' became much more than a revival of the
Classical manner for individual buildings. Architects became
conscious of the importance of the placing of public buildings, in
the hierarchy of building types in the public realm, in town planning
and in the whole question of successful urban design. All this was in
conscious reaction to what was perceived as the chaos, degradation
and unbridled individualism of the Victorian city, and the
movement achieved a triumph with the acceptance of the plan by Sir
Edwin Lutyens for the new city of New Delhi.

In the years leading up to the Great War, the tastes of architects
advanced to embrace the Classicism of the eighteenth century and
the early nineteenth century. The Georgian and Regency periods
became models for urbanity and civic design; the virtues of
repetition, of order, of stylistic consistency became much admired.
This reaction against the architectural individualism of the Victorian
Age was encouraged by the architectural schools which were
supplanting the older training of articled pupilage. It is best seen in
the student work of the Liverpool School of Architecture under
C. H. Reilly, in which a revived Neo-Classical style was applied not
only to public buildings and monuments but also to such things as a
cab shelter.

This movement in design, which was paralleled by the
contemporary reform of typography on Classical lines by Edward
Johnston, Eric Gill and Stanley Morrison, resulted in a public
consciousness of the importance of civic design in the years around

1920. It was a rare and special moment in British history, when both public authorities and industrial concerns strove to commission designs of high quality. Variety and showiness were eschewed in favour of uniformity, reticence and elegance. This manifested itself in the work of the Design and Industries Association, founded in 1915; in the lettering and station designs commissioned for the London Underground by Frank Pick – in particular, the monumental stripped-Classical stations for the Northern Line extension to Morden designed in the early 1920s by Charles Holden; in the consistently high standard of cemetery and memorial design maintained by the Imperial War Graves Commission, established in 1917, under its principal architects, Lutyens, Blomfield, Herbert Baker and Holden; in the authoritative Trajan lettering adopted by the GPO for its dignified and sober new Post Offices designed by the Office of Works in the Classical manner – and, of course, in Sir Giles Scott's telephone kiosk.

Not that Scott's kiosk is a conventional Classical or Neo-Georgian design. The result of a sophisticated architect adapting the Classical language to an unprecedented purpose, the K2 was a distillation of the essence of Classicism. There are no columns or purely decorative details; it relies on proportion, on the appropriate use of mouldings, on the careful placing of projections and set-backs. The most literal element of the design is the saucer dome rising above four segment-headed pediments. It is often assumed that this was derived from the tomb of Sir John Soane in St Pancras Churchyard. Certainly this form is Soanian, as is the use of a reeded Grecian surround to the door and window panels, and Scott was certainly influenced by the revival of interest in Neo-Classicism which encouraged a renewed interest in the long neglected works of the great architect of the Bank of England. It may be significant that in 1925, the year after he designed the kiosk, Scott became a Trustee of Sir John Soane's Museum. On the other hand, it is unlikely that Scott was directly inspired by the dome of Soane's tomb or, indeed, by the similar form of the lantern above the mausoleum at the

Dulwich Picture Gallery. His designs were often subconsciously influenced by a whole range of images he had once seen, and a dome above segmental curves is, in fact, a logical solution to the geometrical problem of designing a sculptural termination to a square pillar when a flat top is not suitable.

The Kiosk No. 2 was, in fact, a miniature building. It is much more than a sentry box, more than a utilitarian structure. Because of the care taken over details and because it is a Classical conception, rising from a podium to a pediment, it has a monumental presence. That might have made it obtrusive in the urban settings in which it was placed. Paradoxically, however, its strong character made it a most sympathetic and respectful neighbour. All the best street furniture has a positive, solid character; it is the negative, purely utilitarian designs which often have the most discordant and distracting presence. The success of Scott's kiosk lay partly in this: in its carefully considered form as a complete, monumental object – as with, say, a bollard or a pillar box. It also lay in the careful and subtle placing of these large objects.

'A SEDAN CHAIR PARK? No – just a nice tidy corner of London's art centre – telephone boxes at the back of Burlington House' ran the caption (possibly written by John Betjeman) to a photograph published in the *Architectural Review* in April 1934.[8] It showed a group of four K2s, carefully placed to line up with each other and with the architecture behind. If a single K2 has a monumental presence, a group of them is a grand architectural statement. Such Classical compositions went well with Classical buildings and, from the beginning, Scott's kiosks were carefully positioned in the streets and were often disposed in pairs to relate to public buildings. In the London of the 1920s – and in London today – Scott's kiosks were as harmonious as they were dignified. Rarely has a public utility had so elegant and urbane an expression.

The K2 rose to some 9 feet 3 inches from the ground and sat on a base 3 feet 6 inches square. It was made entirely of cast-iron sections with the exception of the door, which was of teak, and the floor, which was of granolithic over a base of concrete within the cast-iron frame. One side, which bore the telephone apparatus, was solid iron; the other three had eighteen panes each of 32-ounce sheet glass. Scott made full-sized working drawings for the K2 and designed every detail, including the form of the pierced crown in the pediments (for ventilation) and the lettering of 'Telephone' painted on the 'opal' of frosted glass in the frieze below. This was illuminated by the internal light in the dome at night. The first production K2s appeared in the streets of Kensington and Holborn in 1926 (Scott's wooden prototype was re-erected, with slight modification, within the entrance colonnades of Burlington House off Piccadilly, where it still stands in use). They cost £35.14s.0d each, and most were erected in London. As the Director General of the Post Office noted in 1934, the K2 was expensive and

> is normally erected only in the Metropolitan Boroughs of London. These had refused to allow erection of Kiosk No. 1 and we agreed in their case to use the Kiosk No. 2 (which was designed largely to overcome their objections) ... Elsewhere, Kiosk No. 2 is supplied only in exceptional circumstances – for instance, where the kiosk will be close to the general offices of a great Municipality ...[9]

So a K2 is still to be found next to Liverpool's fine Classical Town Hall. A few found their way abroad: there is a K2 in the Dockyard in Bombay. By 1934, 1,700 had been produced.

The success of and critical acclaim enjoyed by the K2 led the Post Office to commission further variations on the design from Scott. In 1928 he was working on a design for a kiosk in reinforced concrete. This was the K3, which went into production in 1929 (p.54). Sometimes described as a more 'refined' design for important locations, this kiosk had a flatter pediment which was supported on vestigial capitals above the frieze, which framed a ventilation slot. In fact, however, the very nature of concrete made the design less refined, for details like the reeded surround and the raised panel

mouldings were omitted. Possibly it was more refined in its
colouring, for only the window surround and glazing bars were
painted pillar-box red; the body of the kiosk was painted in a light
stone colour. Unfortunately, the K 3 suffered from many of the
problems that beset concrete buildings at the time, in addition to the
fact that the concrete elements often broke while being transported.
In the unkind British climate, the concrete spalled and cracked and
the paint peeled. The K 3 was eventually discontinued, but for seven
years it was the Post Office's standard kiosk for locations outside
London. By 1934, 12,000 had been made. Very few survive in
Britain today, although there are still many in Portugal originally
exported by the Anglo-Portuguese Telephone Company.

However, despite the failings of the K 3, this concrete kiosk, along
with the K 2, was responsible for giving the Post Office's public call
box an authoritative image. 'There are signs that the telephone
Kiosk will soon become as familiar an object in our highways and
byeways as the more historic red pillar box,' announced an editorial
in the *Telegraph and Telephone Journal* in 1933; that is, even before
the advent of the triumphant K 6.

> With its cheerful hue by day and its welcoming bright light at
> night, its promise of ready aid to all in need of rapid
> communication, its form as a friendly figure in the scene whether
> it stands in one of a row in a busy railway station or shopping
> centre, or solitary in a suburban High Street, or enbosomed in
> bushes at the entrance to a park or recreation ground, or – a link
> with the urbane world – conspicuous on the village green, it is
> undoubtedly a persuasive standing advertisement of the telephone
> service.[10]

The K 4 returned to stalwart and serviceable cast-iron, but this
experimental kiosk was also a failure (p. 58). It was first designed in
1925 by the GPO's engineers who adapted and stretched Scott's K 2
to make a miniature automated post office. This monster contained
an external post box and stamp machine in addition to an internal

public telephone. The back wall of the kiosk was thickened to incorporate these additional facilities, which meant that the 'Vermillion Giant' was oblong in plan with two of its segmental pediments stretched. It was thus visually unsatisfactory. But its chief failings were that the rolls of stamps tended to get gummed up inside, while the sudden clunk of the machinery alarmed those inside using the telephone. Production began in 1930 and ceased in 1935, by when fifty of these curiosities had been made. A few survive today.

The K5 was a response to the problems encountered with the concrete K3, but it was purely an experimental kiosk which never entered production. The advantage of concrete construction was its cheapness – the K3 cost only £11 each. The GPO therefore experimented with a new concrete design. In November 1934, the Engineer-in-Chief reported to the Director-General that

> It was recognised some time ago that the finish of the concrete sections of Kiosk No. 3 was not satisfactory and investigation showed that the trouble was chiefly due to the complicated moulds necessitated by following strictly Sir Giles Scott's design. An experimental design which simplified the moulds and manufacturing process was proposed and a few sample kiosks made up.

This was, presumably, the K5 which, claimed the engineers, met with Scott's approval 'from both the structural and artistic points of view'.[11] This new concrete design was priced at a mere £8.19s.6d, but it was never required because the following year Scott came up with his most successful and ubiquitous design: the K6 or 'Jubilee Kiosk' (p.60–61).

In the year of the Silver Jubilee of King George V, a GPO committee was set up to consider both improving the design of the telephone apparatus and the design of a new kiosk to replace the K3 which could be mass-produced for sites all over the country, both urban and rural, as part of a campaign to increase the number of public call boxes – the 'Jubilee Concession Scheme' provided one in every village that had a Post Office. Although it cost more than concrete, the Director-General came to favour cast-iron construction as it allowed more space internally – two inches all round – within the same external configuration. It could also take red paint all over. The trouble with the majestic k2 was that it was too large and too heavy for most sites, as well as too expensive. In March 1935, Scott was therefore asked to design a new cast-iron kiosk which retained the 'best characteristics' of the k2 'but occupying no more pavement space than the concrete type'.[12] He was also asked to consider an improved concrete design, which he apparently did. But it was the cast-iron design which secured official approval. Having been approved by the Royal Fine Art Commission, the k6 or 'Jubilee' design went into production in 1936. In 1935 there had been 19,000 public telephones throughout the country; by 1940 there were 35,000.

As may be seen wherever the two Scott kiosks stand together, the k6 did not differ from the k2 in size alone – standing at 8 feet 3 inches rather than 9 feet 3 inches tall, and weighing in at a mere thirteen and a half hundredweight rather than one and a quarter tons (p.62). Sir Giles both simplified and altered the design in interesting ways. Scott's work had changed in character since 1924. Anxious to evolve an architecture that avoided both the extremes of the Modern Movement and of reactionary traditionalism in the polarised architectural politics of the time, Scott was now applying a more 'modernistic' style to new building types. He was responsible for the external treatment of the Battersea Power Station, with its 'jazz modern' vertical fluting of the brickwork, and for the clean lines of the new Waterloo Bridge. Something of this new sensibility can be seen in the design of the k6.

While the basic configuration of the 'Jubilee' kiosk remained the same as the k2, with the characteristic Soanian dome, the reeded Grecian fluting was removed from the door and window surrounds. Scott retained only the curved projections which gave strength to the prefabricated cast-iron panels, now designed to be bolted together and erected in a day. Another simplification was that the pediment was elided with the frieze, with a ventilation slot placed below the glass 'opal', so that the crown in the pediment could now be embossed rather than pierced. At the bottom, the blank panel was removed from the space between the base and the glazing. The most significant change, however, was to the window treatment. Instead of having regular-sized panes of glass of Georgian proportions, Scott made the panes irregular by moving the vertical glazing bars sideways. This created a broad central light with narrow margin lights. He also actually increased the number of horizontal glazing bars, making the windows eight rather than six panes high. The ostensible excuse for this was to achieve better visibility, but the real reason must be aesthetic. 'A modernistic touch, not over-emphasised, is introduced by the horizontal glazing and the feature furnishes a remarkably fine view from the inside of the kiosk,'[13] reported a Post Office Journal in 1936, in words that sound very much like Scott's own. The result was to give the K6 the horizontal proportions typical of *moderne* architecture of the 1930s.

Painted a bright red, the K6 became a ubiquitous beacon, a symbol of the promise of the Post Office to provide efficient communication from the most remote or bleak spot. 'It was considered essential to standardise design and colour to enable the public to recognise a kiosk,' claimed an official in 1954, 'so that assistance could be obtained quickly in an emergency, whether the fire, police, or ambulance service were required.'[14] Never, surely, has an institution been given such a successful symbolic form and it is surely significant that, decades after the K2 and K6 were designed, and years after their designer's death in 1960, the red telephone box still appears on tourist postcards as an unmistakable icon (p. 102).

The red colour was important. The Post Office had consulted the Royal Fine Art Commission in 1924 about the decorative treatment of its kiosks, and the Commission had endorsed the use of 'Post Office red' as a standard colour, easy to spot and giving an

authoritative and official character. But it was only with the advent
of the K6 that all-red kiosks were to rise in rural as well as urban
areas. There was some opposition to the use of red in areas of great
natural beauty, but it was reported in 1936 that red,

> although not without objections from some quarters on official
> grounds, has nevertheless received the approval of the Royal Fine
> Art Commission and this approval is supported by the Councils
> for the Preservation of Rural England, Wales and Scotland, thus
> placing the Post Office in a particularly strong position to justify
> the choice made.[15]

After the war, however, it was agreed that in areas like the Lake
District, kiosks could be painted battleship grey with only the
glazing bars picked out in red. Exported K6s, of course, could bear
different liveries. In independent Hull they are painted white, while
in the Irish Republic they are green.

However, the GPO's were not the only telephone kiosks to appear
in the British landscape between the wars. There were also the
telephone boxes erected by the two motoring organisations, the
Automobile Association and the Royal Automobile Club, by the
roadside in remote spots. The first AA boxes were erected in 1912.
These not only looked like sentry boxes, they were also called
sentry boxes as they were intended as shelters for patrols on point
duty. This wooden sentry-box type remained the standard, with
variations, for the AA, as for the RAC (p.87). A tougher and more
monumental version of the sentry box, with a low, stepped,
pyramidal roof, was adopted by the Police in 1929. This type of box
is now very rare, and is best known as the camouflage for the time-
travelling 'Tardis' in the television adventures of *Dr Who* (p.84).

Some examples of these police boxes survive in Glasgow but not
in Edinburgh, where the city's police force was under the control of
the Corporation of the City of Edinburgh. Just as the Metropolitan
Police had its own free-standing telephone public call posts, so
Edinburgh's police had its own distinctive boxes. These were

designed in 1931–3 by the City Architect, Ebeneezer J. Macrae, and his assistants, A. Rollo and J. A. Tweedie. They produced a Neo-Classical sentry box of great refinement and beauty which harmonises effortlessly with the Classical architecture of the 'Athens of the North'. It is the only other design for a telephone kiosk (and occasional lock-up) that deserves to be compared with Sir Giles Scott's masterpieces (p.85).

The K6 remained in production until the 1960s, and was not superseded until 1968. In 1954, when there were almost 64,000 kiosks of all types operational throughout the country, the Public Relations Department of the GPO was justifiably confident that the Jubilee kiosk had no peer. 'To provide a high-quality service to the public, a well designed, clean and attractive kiosk, with neat and durable fittings, has been the constant policy of the Post Office and it has not failed in this respect,' claimed George Orchin. 'Much more remains to be done and saturation point seems a long way off, but there is no doubt that our Jubilee Kiosk will stand the test of time and remain a hallmark of successful enterprise and development in the interest of our people.'[16]

Such a policy would be equally valid today, except that the successors to that Post Office have become more interested in ephemeral style than in substance. But style in architecture is always affected by fashion and inevitably changes. By the end of the 1950s it was not surprising that the Jubilee kiosk was beginning to seem dated to many designers. The post-war years had brought the victory of a younger generation of architects, committed to the Modern Movement, over the more traditional and older architects like Scott. It was, therefore, inevitable – and right – that the Post Office should consider commissioning a good new design in the modern spirit. However, instead of holding a new competition, the Post Office in 1959 asked the architect Neville Conder, of Casson & Conder, to come up with a new kiosk.

The result was the K7 (p.78), a kiosk of aluminium with glazing on all four sides and the telephone apparatus attached to one of the

four solid chamfered corners which, alone, were coloured red
externally. A band of metal ran round the kiosk at hip height,
otherwise most of the structure was of glass, which rose to the flat
ceiling. Here was the new aesthetic: light, rectilinear and
insubstantial. Typically, the concrete base was *narrower* than the
box itself, so eschewing the traditional Classical principles which
had governed the design of Scott's kiosks. It was a different
approach to street furniture, clearly inspired by Continental
examples. Instead of monumentality there was transparency: yet this
did not, in fact, mean that the kiosk was any less obtrusive.

But the K7 was a failure. A dozen prototypes were placed in the
streets of London in January 1962. While the public liked such
features as the large door handle, which was easier to grip than
Scott's 'cup' handles, the light aluminium and glass box made theft
and vandalism more attractive and failed to stand up to the English
climate. To quote British Telecom's own history of telephone kiosks,

> One kiosk which made a triumphal debut at the Royal Enchange
> was soon reduced to a streaky grey-black, heavily blistered mess;
> the press cameramen who had crowded around it in January
> would have hardly recognised it, let alone wanted to photograph
> it, a matter of months later. Those searching for a fresh 'look' to
> telephone kiosks for the latter part of the twentieth century
> would have to think again.[17]

The irony is that, since 1984, British Telecom has gone down the
same unsatisfactory path.

In the event, the GPO did manage to come up with a successful
and distinctively modern kiosk. This, the K8, was the result of a
competition held in 1965 in response to a need for two thousand
replacements of the Scott kiosks which were particularly desired by
local authorities with modern town centres. What was required was
a kiosk which was "related to the new house style for public
offices".[18] which was fully enclosed with large areas of glass and
which was cheap to produce and to maintain. Above all, however,

the competing designers had not only to produce a kiosk which could stand up to the weather as well as Scott's kiosks, they also had to cope with a growing problem which was proving to be more destructive than the English climate: vandalism. In short, what was required was a kiosk for the 1960s – a decade less propitious for civic art than the distant and less violent 1920s.

In the event, Conder withdrew from the project and only Martin and Scott competed. Martin submitted a design to be made in aluminium alloy with large panes of glass. Scott, in conjunction with W. Rossi Ashton, proposed a smaller kiosk in cast-iron with a rigid structure whose simple corner joints would be enclosed by a stainless steel trim. Full-sized models were made of each design. In May 1966, the Post-Master General expressed a preference for Martin's design, despite the fact that it was not considered to be sufficiently robust as the corner tubes were separate from the side panels. The constructional system was subsequently redesigned, with the corner tubes incorporated into the panels. It was also decided to make the kiosk in cast-iron, with iron also used for the roof instead of fibreglass. An order for the first thousand kiosks was placed with the Lion Foundry in January 1968 at a cost of £100 each (in 1965 a K6 cost £180, equipped and installed) and the first K8 was erected in Old Palace Yard, Westminster, on 12 July – in the middle of that heady summer of student revolt.

Having consulted the Council of Industrial Design, the Post Office invited three designers to submit designs: Conder, Bruce Martin and Douglas Scott. Martin was an architect who had worked on pre-fabricated construction methods for the celebrated school buildings erected by the Hertfordshire County Council after the war and who was interested in "modular co-ordination". Scott (no relation) was the industrial designer who was responsible for the new universal pay-phone equipment which replaced the old black 'Jubilee Assembly' with Button A and Button B in all the old kiosks designed by his namesake (which, in fact, he saw no reason to replace)'[19]

Like the K7, the K8 was conspicuously 'Modern' with its
narrower base, flat roof and large single panes of toughened glass
(p.80). But in many respects it was still in the well-tried tradition of
Scott's kiosks, for it had a solid back wall, and a distinct 'top'
bearing glass panels with the legend 'Telephone' (Conder's K7 was
to be anonymous, with no place for lettering). Above all, it was of
cast-iron construction, solid and heavy – and it was painted red.[20]
Made of only fifty parts, in contrast to the 400 that made up a K6,
the K8 was a triumph of careful and rational design. By 1983, 11,000
K8s had been produced. As it could stand up to vandalism as well as
any kiosk made by man and as, being solid, it was able to take any
new improved payphone equipment, there was no reason why it
could not have continued to be the new standard British Telephone
kiosk. But, a year later, a change had occurred which put the future
of tens of thousands of serviceable K2s, K6s and even K8s in
jeopardy and which dealt a death blow to the tradition of civilised
urban design for public telephones inaugurated in 1924: the
establishment of British Telecom.

In truth, the sad story of the decline of the British telephone kiosk
had begun in the latter days of the Post Office. As with many once
great and public-spirited institutions, the management of the GPO
became increasingly meretricious and irresponsible as the
organisation itself became ossified in the 1970s. Change became
valued for change's sake and an utterly pointless attempt was made
to alter the livery from red to yellow – a colour favoured by post
offices on the Continent. To begin this frivolous assault on
tradition, two K6s in the Edgware Road were painted yellow. The
reaction of the press and the public was hostile. Clearly, therefore, if
the red boxes could not be repainted, the answer must be to replace
the boxes altogether ...
    Meanwhile many kiosks were being ruined by physical alteration.
During the 1970s, the GPO's engineers began to make unauthorised
alterations to both K6s and K2s. Glazing bars were hacked out and

the pattern of small panes were replaced by large sheets of plate glass, K8 style. This was barbarity, for the changes utterly spoiled the subtlety of Scott's designs. The Post Office was clearly losing its sense of aesthetic importance and public responsibility. But much worse was to come with the divorce of the telephone service from the Post Office in 1984. Although they were running what was still, in effect, a corporate monopoly, the management of the newly privatised organisation called British Telecom were anxious to operate on purely commercial principles. An obsession with marketing, with 'logos', with being up-to-date, with 'corporate image', superseded the old tradition of public service and good design.

In January, 1985, the blow fell: British Telecom announced a '£160 million modernisation for Britain's payphones: Building a Payphone Service for the twenty-first century.' This did not just involve the introduction of new payphone equipment, including the admirable Phonecard, but a holocaust of almost all of Britain's existing telephone kiosks. At a cost of some £35 million, the K2, the K6 and the K8 were to be replaced by kiosks of American design and, for the first 330, of American manufacture. These were little more than glazed boxes of anodised aluminium and stainless steel with a band of yellow plastic either at the top or half way down the sides and with doors and walls that failed to reach the ground. Other kiosks were to be without doors altogether, and some were just canopies. These new designs, claimed British Telecom, 'are cheaper to maintain, more resistant to vandalism, and designed to blend in with any surroundings'.[21]

In fact, the new kiosks blended in with their surroundings all too well. Amidst the cacophony of tawdry, garish junk that increasingly filled Britain's streets by the 1980s, they failed to stand out readily as telephone kiosks while being, at the same time, less sympathetic to their surroundings than Scott's – and Martin's – distinctive and monumental classics of street furniture. But the real objection to the new kiosks was that they were as mediocre as they were cheap –

with a projected life of only fifteen years when K6s were designed to last half a century, while K2s have survived even longer. Had British Telecom behaved responsibly and elicited a good new British design for a telephone box, perhaps through a competition, which secured the approval of the Royal Fine Art Commission, there would have been less opposition to the changes. There would also have been little objection had BT simply replaced old kiosks by new ones when they had come to the end of their useful lives. Instead, it was announced that all the old kiosks would be replaced within a decade. This was both extravagant and unnecessary, as new payphone equipment could be installed in old kiosks as easily as in new ones. But BT's priority was not efficiency, or good design, merely the vain pursuit of a new corporate image.

In 1987, Norman St John Stevas, the Chairman of the Royal Fine Art Commission, could write that

> British Telecom is in real danger of bringing privatisation into disrepute … it is now compounding its acts of omission by a major campaign of destruction, aimed against the traditional British red telephone box … What is as wicked as their destruction is their replacement by shoddy, flimsy, badly designed substitutes, downmarket versions of what one would be likely to find in the Bronx in New York.[22]

Against such accusations that BT has neglected its aesthetic responsibilities, the management have produced no convincing defence other than to repeat that the old kiosks are 'outdated' and 'no longer meet the requirements of our customers'',[23] while the new boxes provide 'more vandal-resistant housings which give easier access for the elderly and disabled.'

In fact, the need of human beings – to stand protected when making a telephone call – has not really changed since 1924, or even 1884. Of course the public will prefer new kiosks if they contain equipment which works when that in a vandalised, neglected and filthy old kiosk is out of action, but that is the fault of British

Telecom, not Sir Giles Scott, whose kiosks can house new equipment as well as the new ones. It may be true that the Scott kiosks, on their bases, are not accessible to the disabled in wheelchairs, but they are preferred by other categories of the disabled, such as the hard of hearing and the partially sighted. The argument about kiosks is not about efficiency, but about the quality of public design, of street furniture.

The opposition to the destruction of Britain's 'traditional' telephone kiosks was widespread. It was led by the Thirties Society and by certain local authorities, notably the City of Westminster and the London Borough of Camden, whose conservation officers had the support of English Heritage. In a way, it was like the campaign against the K1 in the early 1920s all over again – except that this time the fight against an ugly kiosk ended in defeat rather than victory. But it was a hard-fought battle. In February 1985, the Thirties Society wrote to every local authority in the country, asking their conservation officers to negotiate with BT's local managers to secure the retention of old kiosks in sensitive locations. The City of Westminster, in particular, was concerned to keep the K2s and K6s along the principal processional routes as these elegant and dignified objects sit so well with historic buildings.

Later that year, the Thirties Society published a report on the threat to the telephone kiosk, *The British Telephone Box – Take it as Red*, written by Clive Aslet and Alan Powers. This contained many responses from local authorities, the great majority of which were favourable to the Society's case. Unfortunately, there was little any conservation officer could do to impede British Telecom's remorseless replacement programme, especially as local agreements were seldom honoured. A small change in conservation area legislation to encompass street furniture would have allowed conservation officers to protect certain kiosks, but this the Department of the Environment declined to make. As a result, the only possible means of saving kiosks was to have some of them statutorily listed as being of historical or architectural importance –

absurd, perhaps, for utilitarian objects, but not so absurd when it is
considered that the majestic K2s are virtually little buildings.
However, despite the urging of English Heritage, the Department of
the Environment proceeded with painful slowness, as if unwilling to
impede the Government's flagship of privatisation.

Only in 1986, eighteen months after the commencement of British
Telecom's modernisation campaign, did the Department begin
protecting early kiosks. On 6 August Lord Elton, Minister of State
for the Environment, formally listed a rare K3 outside the Parrot
House in London Zoo. Typically, when he attempted to make a call
from this box, the Minister found it to be out of order. Also out of
order was the Department's declared policy for listing kiosks, for
the selection of suitable candidates was to be made in collaboration
with British Telecom. As the Thirties Society commented, 'This is
unique in the listing procedure. In no other case is poacher invited
to act gamekeeper as well'[24] – normally listing is done purely on
merit, without consulting a building's owner.

It was also unsatisfactory that the Department seemed only
interested in early and curious examples, like the surviving K4s in
Warrington and Whitley Bay. Furthermore, because, in England,
listing did not then extend beyond 1939, K6s were not eligible unless
it could be demonstrated that they were installed before that date.
The result was that most of the kiosks listed were in London, the
home of the K2. Distinctive examples, like the squadron of K2s off
Bow Street, Covent Garden, or the symmetrical group of K2s and
K6s behind the Royal Courts of Justice (p.62), were protected, but
most kiosks were defenceless. Only in 1987, when the anomalous
1939 listing limit in England was replaced by the 'thirty-year rule',
could local authorities outside London ask for particular K6s to be
considered for listing. Even with this change their future is bleak,
for the Department of the Environment has a quota for the listing of
kiosks. In the future, it is likely that only about 1,300 Scott kiosks
will survive out of an original 60,000, principally in the London area
and also in places where the local authority has been vigorous in

their defence (See Appendix). As for the completely defenceless K8, dating from 1968, it will probably become extinct.

British Telecom's claim that 'Our kiosk modernisation programme ... has always allowed for the retention of the most worthy red telephone boxes in special locations'[25] has turned out to be worthless. Local agreements made with local authorities have not been honoured, while BT has even applied for listed building consent to remove kiosks which have been listed. Sadly, although desirable in itself and, because of British Telecom's behaviour, necessary, the listing of Scott's telephone kiosks is in fact a defeat for their defenders. This is because the K2 and the K6 will become extraordinary: they will be curiosities to be found in a few historic areas, in effect part of the 'Heritage Industry' for tourists, when the original glory of the Scott kiosks was their ubiquity. They were special because they were not special, for seldom has the ordinary been characterised by such excellence.

It is this that makes their systematic destruction so tragic. No vandalism meted out to a kiosk by an individual has equalled that practised systematically by British Telecom. Sound and serviceable kiosks with years of life in them have been uprooted, usually being damaged or completely smashed in the process. Since 1986, those kiosks that survived removal comparatively unscathed have been sold off at auctions, perhaps to end up as shower cabinets in the United States. Many could and should have been kept in store for re-use in sensitive locations, as the City of Westminster has demanded. But British Telecom knows that there is a market for these distinctive and tangible relics of traditional Britain. This is, perhaps, inevitable. If the ethos of the 1920s was that of responsible public service, then that of Mrs Thatcher's Britain is asset-stripping – the profitable disposal of anything of value in the public realm, regardless of the best and long-term interests of the British people. And the final insult, the ultimate triumph of the marketing mentality, is that the new kiosks are to bear advertisements on their blank metal backs – a complete rejection of the civilised attitude towards public amenities which prevailed earlier this century. It is back to the kiosk as a Parisian poster pylon.

The prodigal idiocy of British Telecom's replacement policy was made all the more extraordinary by the attitude of foreigners to the traditional kiosks. Although the lightweight aluminium boxes may have been based on American and Continental models, the old British boxes were held in high regard abroad. The evidence for this is not only the commercial success of tourist postcards depicting the K2 and K6 and the use of the red kiosk in advertisements abroad to represent Britain (something which BT itself well understood when it used the K6 in its own advertisements on the London Underground), but also the number of K6s which have been exported to the Continent as working telephone kiosks. Towns 'twinned' with British towns have been pleased to erect red kiosks, and there are many in West Germany which have been given by the British army. Several stand in Berlin: in working order and unvandalised. Furthermore, the good sense of the principles which governed the design of Scott's box – solidity and monumentality – have become to be seen as valuable for street furniture on the Continent, particularly in historic cities. New kiosks with pedimented tops have appeared in Vienna, while particularly distinguished new kiosks have been erected in Düsseldorf (p.89). It would seem as if British Telecom is the only body not to appreciate the real merits of Scott's kiosks.

But if the destruction of a ubiquitous functional landmark by British Telecom, for reasons that were specious when not mendacious, was a tragedy, the next event in the strange history of the British telephone kiosk was pure farce. Despite all the money spent and kiosks replaced, the standard of service offered by British Telecom's payphones continued to fail to satisfy OFTEL – the Office of Telecommunications – so that, in 1987, real competition was at last permitted. Mercury Communications were allowed to challenge BT's monopoly in providing a public call box service. And, unlike BT, Mercury behaved responsibly in approaching the difficult question of the design of their proposed kiosks. After taking advice, nine designers were approached and the work of three of them eventually selected.

The first Mercury payphones were unveiled on the concourse of Waterloo Station on 27 July 1988. The press was stunned by the sheer vulgarity of the designs offered: 'Mercury offers kiosk kitsch' was one headline.[26] The three were very different. For indoor sites, Fitch & Co., the commercial design consultants, offered a 'communications totem', which most resembled an old-fashioned petrol pump. Machin Designs Ltd., who manufacture conservatories, produced an ogee-headed glazed canopy that resembled a conservatory. The only true kiosk was designed by the architect John Simpson, a committed Classicist, who proposed a Classical solution: a glazed box with Greek Doric columns at the corners (p.101). For one critic, this was 'not just another piece of trivial postmodern pastiche. It is the monstrous contrivance which finally finishes off the idea of universal public service by reducing it entirely to a matter of style.'[27]

Mercury might well have felt aggrieved at the hostile response to their well-intentioned efforts. The trouble was, in part, the decision to offer three different designs, so denying the merit of universality which the old GPO strove for. As the architectural correspondent of the *Financial Times* complained, 'The telephone box now joins that transitory world of ephemeral fashion, when before it was a distinguished element in all our towns, villages and cities.'[28] But, to a certain extent, Mercury was not well served by its chosen designers. Perhaps it could not be otherwise, for the cultural and artistic civilisation that engendered Sir Giles Scott's design has disappeared. It is instructive to compare the Simpson kiosk with a K2. Whereas in Scott's design, the Classical language is so well digested and understood that only its essence governs the appropriate form, Simpson's kiosk is not only painfully over-elaborate but naively literal. There are no columns in the K2; they are preposterous on such a small structure as a telephone kiosk, but Simpson felt that

Classicism requires complete Doric columns. Yet his Classical elements do not add up to a well composed, harmonious form. The profile is angular and the pedimented top is given a superfluous finial, while the Grecian ornament sits uneasily with the flashy, Superman-style Mercury 'logo'. It is hard to avoid the painful conclusion that, in the chaotic state of modern British architecture, no designer is capable of producing a telephone kiosk as assured and sophisticated as Scott's.

The Royal Fine Art Commission has criticised all three Mercury designs, and new versions may appear. It is also possible that the company may use the moulds kept by the Carron Cabinet and Joinery Manufacturing Co Ltd in Falkirk to cast new examples of the κ6 for their own use (shorn of the Royal crown, of course, as in Hull). If this happens, it will be gratifying as proof of the continuing transcendent excellence of Sir Giles Scott's design, even if the old virtue of authoritative uniformity in the cause of public service will be denied by the wide variety of different kiosks in use, all in different liveries.

The defence of the Scott telephone kiosk was not an exercise in nostalgia, a sentimental protest at yet another change in the public face of Great Britain and at the extinction of a harmless and enjoyable survival from the past. For the κ2s and κ6s were not only admirable and efficient as weathertight receptacles for public telephones: they were also supremely excellent models of how sensitive, dignified and, yes, *beautiful* street furniture can be. This is important. The utilitarian need not be ugly. Those responsible for the appearance of objects in the public realm – above all, those concerned with the housing of public telephones – would do well to reflect on the wise words of Alexander Graham Bell: 'We are all too much inclined, I think, to walk through life with our eyes closed. There are things around us and right at our very feet that we have never seen, because we have never really looked.'[29] For Bell was, after all, the inventor of the telephone.

A true kiosk: an attended public call office in High Holborn *c.* 1903 (*Telecom Technology Showcase*).

above left   An Edwardian 'Norwich'-type telephone box (*Telecom Technology Showcase*).

above right   A public call office in the rustic style outside Blackburn, 1907 (*Telecom Technology Showcase*).

A rural Kiosk No. 1 in the Lake District (*Telecom Technology Showcase*).

A KI of the concrete variety on the northern approach to Southwark Bridge, 1924. Note the Police telephone stand on the right (*Telecom Technology Showcase*).

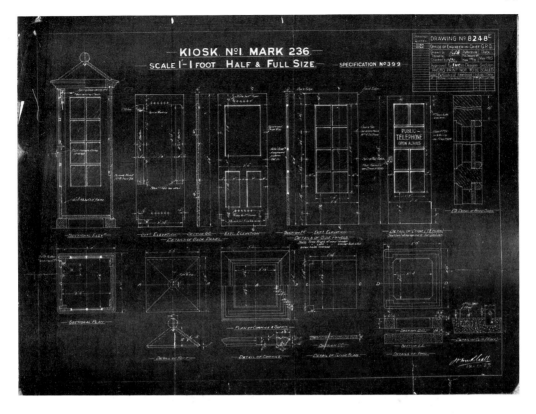

Working drawings of a KI, Mark 236, of 1923, revised 1927, which were sent to Sir Giles Scott (*British Architectural Library*).

'This Arshetecture': thatched K1s in Eastbourne in 1935 as illustrated in *The Architects' Journal*

A monumental curiosity: a stone kiosk in Pittencrieff, Dunfermline, in 1938 (*Telecom Technology Showcase*).

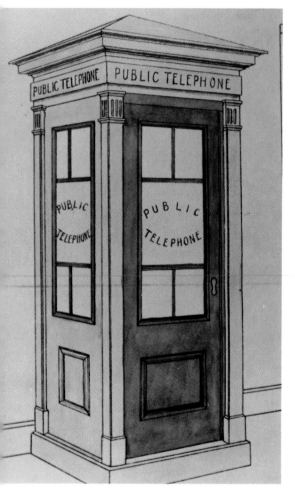 

Designs for kiosks submitted in the 1923 competition organized by
the Metropolitan Boroughs Joint Standing Committee (*Telecom
Technology Showcase*).

C. F. A. Voysey's alternative designs for a Gothic kiosk for Vickers 1923. (*The Builder*, 1925, and *Telecom Technology Showcase*).

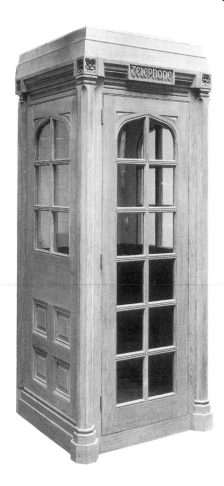

An unidentified Gothic kiosk in wood (*Telecom Technology Showcase*).

Swedish grace? For comparison, a new design of municipal
telephone at Gothenburg in 1924. The open lattice work was to
allow ventilation (*The Hulton Picture Library*).

above left   Full-sized model of the GPO's proposed kiosk design, 1924 (*Telecom Technology Showcase*).

above right   Full-sized model of the Birmingham Civic Society's design revised for cast-iron construction (*Telecom Technology Showcase*).

above left   Full-sized model of Sir Robert Lorimer's design for a kiosk, 1924.

above right   Full-sized model of Sir John Burnet's lampshade design, 1924 (*Telecom Technology Showcase*).

above Scott's original design for the telephone kiosk submitted in
the 1924 competition (*British Architectural Library*).

facing page Full-sized model of Sir Giles Scott's winning design,
1924, now placed at the entrance to Burlington House (*Telecom
Technology Showcase*).

above left   The Soane Tomb, originally designed for Mr Soane in 1816 by Sir John Soane (1753–1837), in the former St Giles's Burial Ground near Old St Pancras Church.

above right   The Soanian top of Scott's K2.

facing page   Scott's own copy of the sheet of full-sized details of the K2 (*British Architectural Library*).

DESIGN FOR TELEPHONE KIOSK
FULL SIZE DETAILS

TELEP

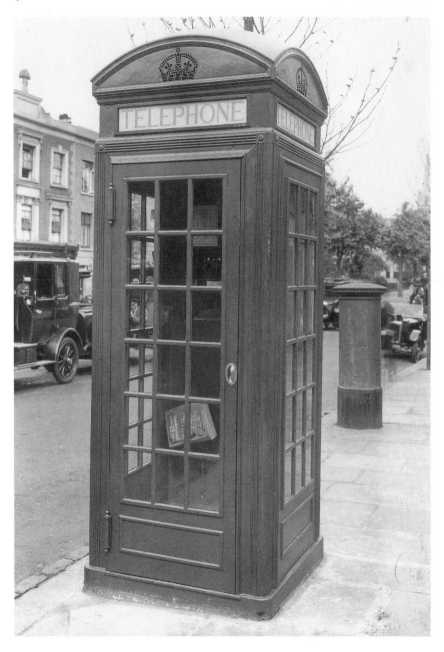

facing page One of the first production K2s in Ladbroke Grove in 1926 (*National Monuments Record*).

above A K2 (with blocked-in glazing) across the Thames from the same designer's Battersea Power Station.

A K2 at the City end of Blackfriars Bridge in 1987 with a Police
Public Call Post nearby.

Two well-sited K2s in Dalston Lane in 1986, since removed.

The author admiring the elegance of a single κ2 well placed below
the portico of Hawksmoor's Christ Church Spitalfields in 1977,
since taken away.

The K2 outside Liverpool's neo-Classical Town Hall.

K2s on Clerkenwell Green in 1984.

Close friends: a pair of suburban K2s in North End Lane.

The first kiosk to be listed: a concrete κ3 under the eaves of the
Parrot House in London Zoo in 1986.

A stage-prop version of the κ3 erected in King's Cross for the
filming of *The Ladykillers* in 1955 with Alec Guinness and Herbert
Lom and somebody else sharing a call (*National Film Archive,
London*).

Giles Scott meets Le Corbusier: an Italian K2 underneath the
museum in Allahabad (*Dan Cruickshank*).

The outposts of Empire: a K2 still on duty in Wellington, New
Zealand, in 1988 (*Matthew Tierney*).

The Warrington K4: a rare listed Vermillion Giant with the Post
Box still operational.

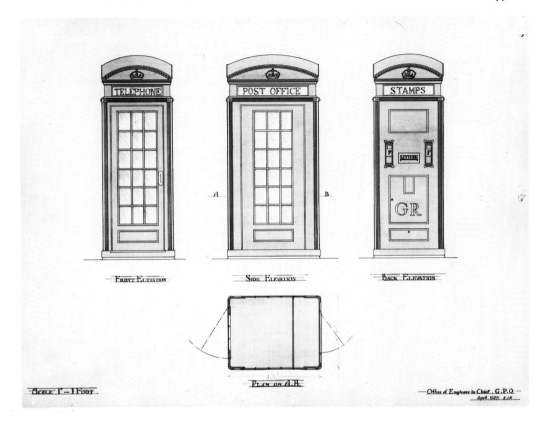

The GPO's design for adapting the K2 to incorporate an automated
Post Office, 1925 (*Post Office*).

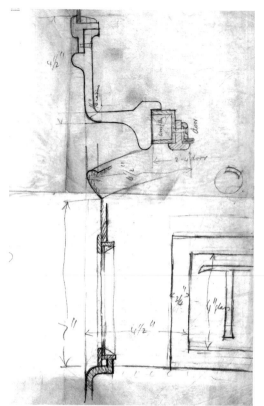

above left   Original sketch by Scott for the Jubilee Kiosk, or κ6, May 1935 (*Lesslie K. Watson*).

above right   Perspective sketch by Scott for the κ6 (*British Architectural Library*).

A GPO poster showing the new Jubilee Kiosk, 1936 (*Telecom Technology Showcase*).

The Happy Family of K2s and K6s behind the Law Courts in Carey Street.

above left   A prototype K6 showing off its 'Jubilee Assembly' (*Telecom Technology Showcase*).

above right   Inside a K6 in Welwyn Garden City in 1986.

above left   K6s nestling in the Edwardian rustication of the County Hall.

above right   A pair of K6s outside Torquay Town Hall.

A lonely к6 maintaining a presence in Bow Common in 1986.

Kiosk meets kiosk in Torquay.

top   A smart seaside K6 at Port Erin on the Isle of Man.

above Outpost of civilization: a solitary K6 on a distant tip of the British mainland on the Ardnamurchan peninsula, Scotland (*Celina Fox*).

above left   Kiosks and obelisk, Ramsgate, 1985.

above right   Braving the elements: a pair of K6s in Brooke Market, Holborn, in the Winter of 1985, since removed.

facing page, top   K6s carefully placed outside Bolton's contemporary 1930s Classical Post Office in 1988.

facing page, bottom   A disciplined squadron of K6s on guard outside Preston Head Post Office in 1986.

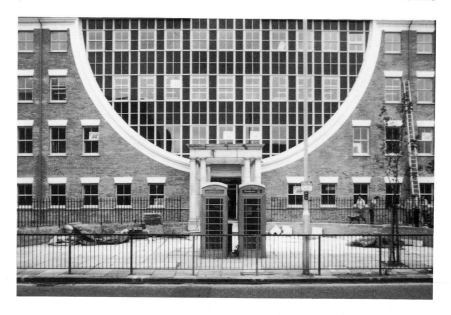

above   Pre-existing κ6s – since replaced – enhancing the
architecture of Campbell, Zoglovitch, Wilkinson & Gough in the
Mile End Road in 1986.

facing page, top   Carefully sited κ6s and a pillar box placed outside
the contemporary Neo-Georgian Post Office in Tenterden in 1987.

facing page, bottom   A defensive square of κ6s in Stafford.

A pair of K6s on guard duty outside Hyde Park in 1987.

above left   A K6 on the Victorian Embankment by Hungerford Bridge.

above right   A discreet K6 near Bexhill in 1987.

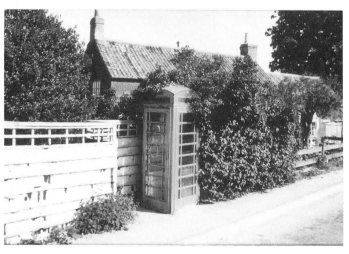

facing page, top A rural K6 at Pickenham in Norfolk in 1986.

facing page, bottom A K6 at Old Sowerby, Lincolnshire, becoming overwhelmed by its surroundings.

above A suburban K6 in Welwyn Garden City in 1986.

facing page, top left   Hull's crown-less K6s in their smart white livery.

facing page, top right   K6 in environmentally sensitive colouring at Cockington, Devon.

facing page, bottom   A phalanx of K6s in dress uniform in Park Lane (*Philip Davies*).

above left   Elite K6s in conservation area uniform in Chester.

above right   A deferential K6 in heritage livery by Royal Crescent in Bath.

One of Neville Conder's prototype κ7s behind the Royal Exchange in 1962 (*Telecom Technology Showcase*).

The full-sized model of Douglas Scott's competition design for a new kiosk, 1965 (*Douglas Scott*).

One of the first production models of Bruce Martin's winning design for the K8 placed outside the GPO in King Edward Street in 1968 (*Telecom Technology Showcase*).

above   A pair of K8s erected at the entrance to Lord Holford's new Paternoster Square redevelopment.

overleaf   A formidable detachment of K8s occupying the centre of Bristol.

A Police Box in Glasgow of the sort employed by Dr Who for
time-travelling.

One of Edinburgh's elegant and special Police Boxes placed in the
Grassmarket with its windows with diagonal cross glazing bars
blocked in.

above left   One of the first of the new Police Public Call Posts
erected in Piccadilly Circus in 1937 (*The Hulton Picture Library*).

above right   A Police Public Call Post at the boundary of the City
of London in High Holborn.

An Automobile Association sentry box of the 1956 pattern but shorn of its original signs (*Philip Davies*).

Belgian telephone boxes outside the main Post Office in Ostend.

above left   Telephone boxes in Bucharest.

above right   A new German telephone kiosk of superior design in
the Königsallee in Dusseldorf.

above and facing   K6s in King's Cross, replaced by British
Telecom's new 'housings', 1986–7.

The sort of flimsy new box BT chooses to display outside its
headquarters in Glasgow.

Decline and fall? A κ6, a κ8 and one of BT's replacements outside
Christ Church, Spitalfields in 1986. The two red boxes have since
been liquidated.

Victor and vanquished: a κ6 and its threatened replacement.

The new BT American-style boxes in Oxford Street in 1985 (*Philip Davies*).

Declining standards: BT neglects to paint the kiosk but the attached
Post Box is still Pillar Box Red as both ought to be (*James
Dunnett*).

K2s off the Lower Clapton Road with stickers to show they have been listed – so why has BT removed the teak door from the one behind?

Who is the real vandal?

Charles Moore, Editor of the *Spectator*, laments the death of a
friend in Amwell Street, 1987.

Mercury's answer: Conservatory-style 'pylons' by Machin Design
at Waterloo Station, 1988 (*Mercury Communications*).

above left   Another answer: a 'communications totem' by Fitch & Co at Waterloo Station, 1988 (*Mercury Communications*).

above right   Battle of the Styles: John Simpson's Classical kiosk for Mercury 'designed to be part of the traditional urban townscape'. (*Mercury Communications*).

A Symbol of Britain: tourist postcards of traditional Scott kiosks
(From *Whiteway Publications Ltd, Fincom Holdings Ltd,
Kardorama Ltd,* and *Real Ireland Design Ltd*).

APPENDIX

Listed Kiosks

By December 1988, the following telephone kiosks had been listed:

K1s – 5 (High Street, Bembridge, Isle of Wight; Market Hall,
  Trinity House Lane, Kingston-upon-Hull; Crich Tramway
  Museum, Derbyshire; outside Boulton's Lock Hotel, Ray Mill
  Island, Maidenhead, Berks; East Anglia Transport Museum,
  Carlton Colville, Suffolk).

K2s – 216 (213 in Greater London and County Court, King Street,
  Shorne, Kent; outside Liverpool Town Hall; Bradshaw Lane,
  Bradshaw, West Yorkshire).

K3s – 2 (Parrot House, London Zoo; Chobham Bus Museum,
  Elmbridge, Surrey).

K4s – 5 (Bridge Foot, Warrington; Church Street, Frodsham,
  Cheshire; Tunstall caravan park, near Roos, Humberside; Station
  Road, Whitley Bay, Tyne and Wear, Northumberland; Dinting
  Railway Centre, Glossop, Derbyshire).

K6s – 996.

Police Call Boxes – 8 (City of London: Queen Victoria Street;
  Guildhall Yard; Hounsditch/Aldgate; Adams Court, Old Broad
  Street; Walbrook; Victoria Embankment. Westminster: Piccadilly
  Circus; Grosvenor Square).

Grand Total – 1,232.

REFERENCES

1. Illustrated as 'This Arshetecture' in the *Architect's Journal*, 16 May 1935, p. 759.
2. *The Builder*, 1 May 1925, p. 664. Voysey's design is preserved in the RIBA Drawings Collection.
3. *Architects' Journal*, 6 February 1924, p. 265.
4. *Architects' Journal*, 8 April 1925, p. 568. The entries in the 1923 competition remain to a certain extent mysterious. A memorandum in the Post Office archives notes that the following nine designs were examined by the Fine Art Commission in April 1924: A) by Messrs Goslin & Sons, put forward by the Metropolitan Boroughs Joint Standing Committee; B) by Morris Westminster Metalworks, obtained by MBJSC; C) & D) by Vickers Ltd., obtained by MBJSC [i.e. by C. F. A. Voysey]; E) by Messrs D. G. Somerville & Co. (re-inforced concrete), obtained by the Post Office; I) Design of Mr Tite, Inspector of Post Office Engineering Department; J) by the Birmingham Civic Society, submitted by the Post Office; K) by H. M. Office of Works; L) Design put forward by the British Institute of Industrial Art after being invited to offer their advice in Design A. Designs F and G were not listed.
5. Sir Robert Lorimer's report accompanying his design, 22 May 1924 (Post Office archives).
6. Scott's report, n.d. (Post Office archives).
7. *Architect & Building News*, 10 April 1925, p. 280.
8. *Architectural Review*, April 1934. This was attributed to John Betjeman by Bevis Hillier in *Young Betjeman* (1988).
9. Memorandum from the Director-General of the G.P.O., 24 September 1934 (Post Office archives).
10. *The Telegraph and Telephone Journal*, July 1933, p. 222.
11. Memorandum from the Engineer-in-Chief to the Director-General, GPO, November 1934 (Post Office archives). However, according to issue No. 2 of *Telecom Heritage*, 1987, p. 36, 'the K5 was a collapsible unit made from armour-

ply (a metal-faced plywood). It was designed for use at exhibitions or trade fairs where temporary facilities were required.'

12. *Builder*, 13 March 1936, p. 522.

13. F. J. Judd, 'Kiosks', *The Post Office Electrical Engineers' Journal*, October 1936, p. 175.

14. George Orchin, Public Relations Department, 'You may telephone from here', *Post Office Telecommunications Journal*, May 1954, p. 110.

15. F. J. Judd, 'Kiosks', *The Post Office Electrical Engineers' Journal*, October 1936, p. 175.

16. Orchin, *op. cit.*, p. 110. Scott apparently was paid £1 for every K6 produced. Bruce Martin, designer of the K8, was not so well treated by the GPO and received only a flat fee.

17. *Britain's Public Payphones*, British Telecom, 1984, p. 22.

18. P.O. report on No. 8 Kiosk by R. W. A. Welch, August 1968 (Douglas Scott).

19. Jonathan Glancey, *Douglas Scott*, Design Council, 1988. Martin found that the K6 met every demand of the GPO's brief except that it was difficult to clean inside and was deficient in ventilation.

20. Martin successfully resisted the proposal by the GPO's design consultants, Henrion, to have the K8 painted yellow. The door of the K8 was of aluminium. Martin, who had experience of aircraft manufacture during the Second World War, originally designed the whole kiosk to allow for the latest French technology in casting aluminium. He also conceived the K8 entirely in metric measurements. The lettering in the opals was drawn out by Maurice Goldring. Martin had hoped to have the Royal Cipher cast in the panel below the window of the door: this was vetoed, probably by the Postmaster General who, in the period 1964–66, was Anthony Wedgwood-Benn.

21. British Telecom News Release, 17 January 1985.

22. Norman St John Stevas, 'Holding a thin red line', *Daily Telegraph*, 7 September 1987.

23. British Telecom News Releases, 17 January 1985 and 18 May 1987.

24. Thirties Society Press Release, No. 37, August 1986.

25. British Telecom News Release, 18 May 1987.

26. *Building Design*, 29 July 1988, p. 36.

27. Patrick Wright, 'On a Ring and a Prayer', *New Statesman and Society*, 5 August 1988, p. 23.

28. Colin Amery, *Financial Times*, 8 August 1988.

29. As quoted on a British Telecom bookmark.

BIBLIOGRAPHY

a) Historical

George Orchin, ‘“You may telephone from here”: The story of the public call office’ in *Post Office Telecommunications Journal*, May 1954, pp. 105–110.

(M. Goss), *Britain's Public Payphones*, British Telecom, 1984.

Andrew Emerson, *Old Telephones*, Shire Publications, 1986.

Publications of the Telecommunications Heritage Group (c/o Telecom Technology, Showcase, 135 Queen Victoria Street, London EC4V 4AT) beginning January 1987.

b) The Battle of the Box

‘Piloti’, “Bugger Buzby” in ‘Nooks and Corners’, *Private Eye* no. 497, 2 January 1981, p. 7.

Philip Davies, ‘The Changing Telephone Box’ in the *Illustrated London News*, May 1985, pp. 21–23.

Clive Aslet and Alan Powers, *The British Telephone Box . . . take it as red*, Thirties Society report, 1985.

Gavin Stamp, ‘Save the Telephone Box’ in *The Spectator*, 9 February 1985, pp. 12–13. Reprinted as ‘The Battle of the Phone Booth’ in *Designer*, April 1985, pp. 16–18.

Charles Moore, ‘Better red than dead’ in *Daily Telegraph*, 29 December 1986.

Norman St John Stevas, ‘Holding a thin red line’, in the *Daily Telegraph*, 7 September 1987.

Clive Aslet, ‘Ring out the Old’, in *Country Life*, 30 April 1987, pp. 118–119.

Patrick Wright, ‘On a ring and a prayer’ in *New Statesman & Society*, 5 August 1988, pp. 21–23.

Peter Kellner, ‘Symptom of a flawed vision’ in *Independent*, 15 August 1988.